Limekiln in operation, perhaps in Scotland, showing a loading arrangement powered by a horse-gin. From a painting by Charles Towne, previously attributed to Patrick Nasmyth.

LIMEKILNS AND LIMEBURNING

Richard Williams

Shire Publications Ltd

CONTENTS

The history of limeburning 3

Chemistry and agrochemistry 8

Kiln structure 11

Transport and loading arrangements 27

Further reading 31

Places to visit 32

Copyright © 1989 by Richard Williams. First published 1989. Shire Album 236. ISBN 0 7478 0037 5.
All rights reserved. No part of this publication may be reproduced or transmitted in any form or by any means, electronic or mechanical, including photocopy, recording, or any information storage and retrieval system, without permission in writing from the publishers, Shire Publications Ltd, Cromwell House, Church Street, Princes Risborough, Aylesbury, Bucks HP17 9AJ, UK.

Printed in Great Britain by C. I. Thomas & Sons (Haverfordwest) Ltd, Press Buildings, Merlins Bridge, Haverfordwest, Dyfed.

British Library Cataloguing in Publication Data: Williams, Richard. Limekilns and Limeburning. 1. Great Britain. Limeburning industries, History. I. Title. 338. 4'766693'0941. ISBN 0-7478-0037-5.

ACKNOWLEDGEMENTS
  The author wishes to acknowledge the help of many people who have provided useful information, and in particular Richard Clarke of Closeburn, Dumfriesshire; John Creasey of the Museum of English Rural Life, Reading, Berkshire; Cynthia Gaskell Brown of Plymouth City Museum, Devon; Brian Dix of Northampton Archaeology Unit; W. N. Pilling of ICI, Buxton, Derbyshire; Michael Nix of Hartland Quay Museum, Devon; and B. C. Skinner of the University of Edinburgh. Illustrations are acknowledged as follows: Ashmolean Museum, Oxford, page 17; Dr D. W. Boyd, page 1; the Black Country Museum Trust, page 28; British Museum, page 23; Richard J. Clarke and Dumfriesshire and Galloway Natural History Society, page 30 (upper); Derby Industrial Museum, page 16; Brian Dix and the *Oxford Journal of Archaeology*, page 3; Eric Holden and the Sussex Archaeological Society, page 5 (upper); ICI, pages 26, 29, 32; Martin Watts and Peter Stanier, page 30 (lower). The cover photograph is by Cadbury Lamb. All the rest are by the author.

Cover: *Eighteenth-century limekilns at Beadnell Harbour, Northumberland (National Trust)*.

Below: *Limekiln in Aberystwyth harbour, Dyfed. The sites of as many as fourteen kilns are known around this harbour but the structures of only three survive. In the 1860s fifteen ships regularly brought in limestone and others landed culm as fuel.*

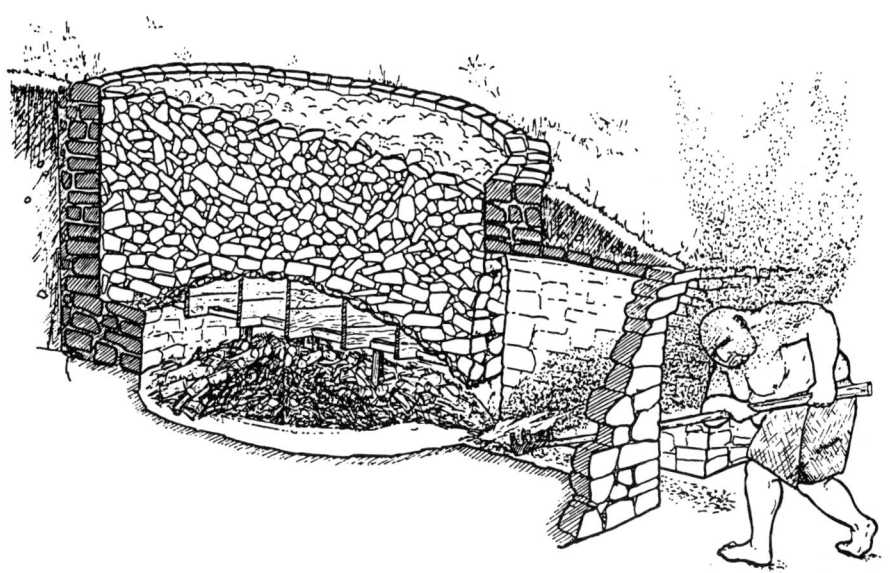

Conjectural drawing of a Roman periodic or flare kiln, based on excavations. The initial charge was formed into a dome resting on an internal ledge and supported by a wooden framework. The fuel was wood.

THE HISTORY OF LIMEBURNING

'Lime is an article of great consideration,' said C. Gray in W. H. Pyne's *Microcosm*, published in 1808, 'not only from its utility for various purposes, but from the employment which the manufacturing of it affords to thousands. From the quantity used in building houses, walls, plastering, &c. it forms an important item in the national expenditure. The consumption of it has been also of late greatly increased by its successful application to agricultural purposes, by our improving farmers.' He might have added that in Britain it was also used, in the form of a lime-wash, to waterproof walls and to lighten interiors; that lime water was taken as a medicine; and that lime was used for bleaching paper and for the preparation of hides in leather-making. A more gruesome use was practised in parts of the Far East; bodies were buried in lime and when the flesh disintegrated the bones would be washed and placed in pots on hillsides. In the Canary Islands, and no doubt elsewhere, bodies were limed at burial for hygienic reasons. Today, lime is still used in agriculture and horticulture and there are extensive applications in the chemical industry, in water purification and in effluent treatment.

Evidence of limeburning in prehistoric times was found when a limekiln dating from about 2450 BC was excavated at Khafaje, Mesopotamia. Mortar made from lime was used in Crete in the Middle Minoan period, about 1800 BC. The use of lime in early prehistory was not widespread, however; buildings were often of mud-brick, and where there was masonry it was either unbonded or other substances were used. For example, the Ishtar Gate of Babylon used facing bricks bedded in bitumen. The ancient Egyptians used gypsum plaster in pyramids as gypsum can be burnt at a much lower temperature than limestone. The necessary quantity of wood as fuel for limeburning was not readily available although the limestone was abundant.

In classical times, the Greeks used lime mortar in pavements and aqueducts, but the extensive use of lime and the development of well built kilns must be attributed

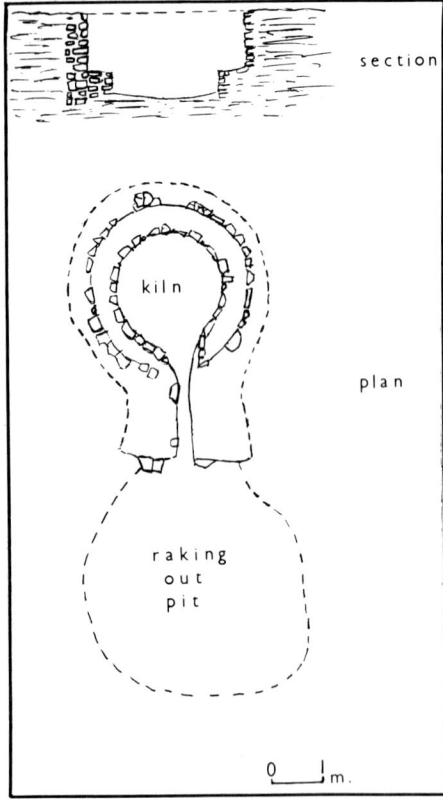

Roman limekiln at Weekley, Northamptonshire, from excavation drawings. The pot was about 10 feet (3 metres) in diameter and had an internal ledge or bench. It was built of coursed limestone backed by rubble and survived to a height of about 6 feet 6 inches (2 metres). Note the long stoke-hole and raking-out pit. (Based on a drawing in the excavation report of D. A. Jackson in 'Britannia', the journal of the Society for Promotion of Roman Studies.)

to those great engineers, the Romans. The Greeks and Romans even learnt how to make strong water-resistant cement and the Romans developed a form of concrete. The value of lime in agriculture also appears to have been first appreciated by the Romans although, where available, natural calcareous substances such as marl were applied to sour land. Pliny referred to the use of 'white chalky marl' in Britain. The earliest description of a limekiln is in Cato's *De Agricultura*, written in the second century BC, a book giving advice on cultivation and land management.

In Britain, the earliest archaeological evidence for limeburning is also found in the Roman period. It is surprising that so few kilns have yet been excavated at Roman sites since extensive use of mortar and cement is apparent in surviving structures. A battery of six limekilns excavated at a legionary site at Iversheim, West Germany, shows that they produced lime in quantity on military sites.

Following the Roman period, there was very little demand for mortar in Britain until the founding of the great religious houses and castles. However, Saxon references to lime and mortar, and a twelfth-century statement that the church at York had been whitewashed with lime in AD 690, indicate that limeburning was practised to some extent. Lime was often made in pits or clamps using a method akin to charcoal burning, leaving little evidence. In medieval times, however, limekilns are specifically mentioned in manuscripts and have been confirmed by excavation.

Remains of limekilns have been found at several thirteenth-century castles, and early building contracts included provision for building kilns for on-site production of mortar. There are records of large quantities of timber being burnt in royal limekilns at, for example, Oxford in 1229 and Wellington Forest, Shropshire, in 1255. At the Tower of London in 1278 and at Windsor Castle in 1366, limekilns used coal from the Newcastle upon Tyne area, although the use of coal was prohibited in London in 1307 because of the air pollution (a prohibition which was probably widely ignored). The contract for building a bridge over the Swale at Catterick, North Yorkshire, in 1421 required the builder to excavate sand and limestone and to build limekilns. One of the earliest excavated examples of a medieval kiln was at Guildford, Surrey, in 1984, showing evidence of firing in the twelfth century.

In the Tudor period, large quantities of lime mortar were made at the sites of the great royal palaces. For example, 890

Right: *Fifteenth-century limekiln excavated at Old Erringham, Shoreham, West Sussex. Note the two draw-holes and clay lining. Built for making lime mortar for a manor house, it was about 6 feet (1.8 metres) deep.*

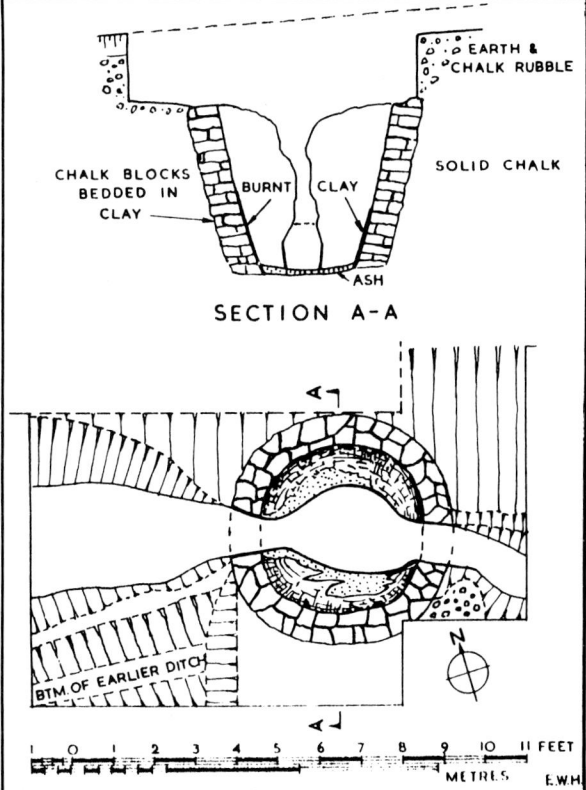

Below: *Thirteenth-century limekiln at Cilgerran Castle, Dyfed. Only the base survives, with two draw-holes. It was coal-burning and was possibly an early type of semi-continuous kiln.*

Kiln at the small harbour of Clovelly, Devon (centre background). The ramp to the kiln top can be seen to the right of the kiln.

Rim of the kiln at Clovelly. Beyond the boat is the working area for loading the kiln.

loads of lime were burnt just between May and September 1538 for Henry VIII's Nonsuch Palace in Surrey, apparently in brick limekilns, using chalk from a nearby pit probably opened for the purpose.

Specific references to the use of lime as a soil improver are found in the early sixteenth century, with implications that the practice started in the fifteenth. It is probable that previously only raw chalk, marl and other natural calcareous materials such as sea-sand had been applied in Britain, and the use of these continued in parallel with burnt lime. Possibly the earliest description in Britain of the construction of a kiln for agricultural lime is in George Owen's *Description of Pembrokeshire* completed in 1603. Enclosed fields were beginning to replace the old method of strip cultivation in open fields and rural limeburning became more common in the seventeenth century.

Simultaneously, the building of large houses and the increased use of brick in vernacular houses, with the addition of chimneys, led to a greater demand for lime mortar. Kilns were built in limestone quarries and chalkpits and the trade of limeburner appears more frequently in documents as a specialised occupation rather than a part-time task of a builder or general labourer.

Until the middle of the eighteenth century limekilns were often temporary structures built solely to meet an immediate demand and then allowed to collapse or were partly rebuilt for the next firing. In some areas they burnt lime in clamps or in *pye kilns* which burnt a quantity of limestone for a week or so and were then dismantled to extract the quicklime. Where good building stone was to hand, a well built kiln could be re-used several times before requiring substantial repair or rebuilding. The more robust kilns still standing intact today, however, are no older than the late eighteenth century and many are of nineteenth-century date.

The agrarian revolution of the eighteenth century, when large areas of farmland were enclosed, created an enormous demand for lime. Landowners cashed in on increasing grain prices in the Napoleonic War years: it has been estimated that 2 million acres (800,000 hectares) of land in Britain were brought into cultivation between 1790 and 1810. Vast numbers of limekilns were built and in some areas every farmer had his own kiln. In coastal areas the many kilns in harbours or beside beaches were sturdily built and capable of lasting for years. Limestone and *culm* (anthracite dust) or slack coal as fuel were shipped in, becoming a major coastal trade. Local farmers bought lime at the kiln or rented a kiln to burn their own. Some were built in massive blocks of two or more draw kilns.

Similarly, blocks or batteries of kilns were built beside canals as the expansion of the waterway network allowed the raw materials for lime production to be carried long distances from their sources.

Battery of coastal kilns near Llanrhystud, Dyfed. The walls are of rubble and there are multiple draw-holes. They stand on a low cliff and remains of a quay are visible on the shore below.

Some canals were even cut specifically for the lime trade. Thus farmland previously deprived of lime could be brought under cultivation. Simultaneously, the expansion of industrial towns and the building of canal locks, bridges, wharves, aqueducts and docks led to an ever increasing demand for mortar.

Thus kilns capable of continuous production became a necessity and kiln-blocks were often of elaborate construction; many innovative designs to increase efficiency and output appeared. By the twentieth century mass production had become concentrated at the larger limestone quarries and chalkpits, the railway network permitting economical distribution from a smaller number of centres.

The introduction of Portland cement reduced the demand for simple lime mortar, while finely ground chalk and limestone became available at competitive prices as an alternative to quicklime for liming the land. Isolated rural lime-kilns had had their day and hundreds, if not thousands, were destroyed or fell into ruin; only in remote areas and in some harbours and estuaries were they operated into the twentieth century. A government subsidy on agricultural lime in 1937 led to the revival and reopening of some disused kilns. For example, draw kilns of the old type were rebuilt at Duncton, West Sussex, in 1940 and one, though uneconomic, was still in use in 1984 alongside more modern kilns.

Modern methods of lime production for agriculture, industry, chemicals, water purification and so on are beyond the scope of this book, and cement manufacture is a subject of its own. Mechanised quarries with horizontal and rotary kilns, together with modern designs of vertical shaft kilns, are found at the small number of production centres. Some are oil- or gas-fired, although coal is making a revival as fuel prices fluctuate.

CHEMISTRY AND AGROCHEMISTRY

Limestone and chalk are forms of calcium carbonate, $CaCO_3$, geologically formed under water from deposits of shells and the skeletons of marine organisms. Carboniferous and Jurassic limestones outcrop over wide areas of England and Ireland, South Wales and the Scottish Lowlands. Cretaceous chalk occurs in the south-east of England, in a belt from Dorset to East Anglia, and each side of the Humber. Open-cast extraction was general but in parts of the South-east, particularly Kent, chalk was mined from underground galleries.

When calcium carbonate is heated to a temperature of 900 to 1100 C (1652 to 2012 F) carbon dioxide, CO_2, is driven off as gas to leave calcium oxide, CaO, known as *quicklime*. This process is called *calcining*. Quicklime reacts exothermically with water, so fiercely as to generate steam, to form calcium hydroxide, $Ca(OH)_2$, known as *slaked* or *hydrated* lime. This slowly reacts further with the carbon dioxide and water of the atmosphere to revert to calcium carbonate, thus completing a cycle.

This property of reverting to carbonate allows lime to be used as a cement. The quicklime is slaked with an excess of water to form lime putty. Mixed with sand, this forms mortar which sets hard as the conversion to carbonate occurs. It is a slow process and the mortar is initially weak, though continuing to harden for years. A harder cement results if the lime contains alumina and silica, which form calcium aluminate and silicate as the cement hardens. These insoluble compounds allow the cement to harden under water, a valuable property for building, and the cement is said to be *hydraulic*. The greystone of the Lower Chalk produces *semi-hydraulic* lime but the best hydraulic limes are obtained from Lias limestone of the Lower Jurassic, occurring in layers with shale, and from shaly Carboniferous limestone.

The Romans added *pozzolana,* a volcanic earth, to lime for a semi-hydraulic cement, and even crushed pottery or tile was beneficial — volcanic clay or tufa was better. Parker's so-called 'Roman cement' in 1796 was made by calcining nodules of argillaceous (clayey, containing alumina) limestone called *septaria* found in the Thames estuary. This was superseded by Portland cement patented in 1824 by Joseph Aspdin, made by burning clay with limestone at a high temperature, the resulting clinker being ground to a powder.

Most of the quicklime produced until the late nineteenth century was, however, used in agriculture. The term *liming* refers generally to the application of calcium compounds to the soil, whether in the form of carbonate, hydroxide or oxide. Calcium is essential for plant growth but the principal effect of liming is to neutralise soil acidity, most crops requiring a neutral or slightly alkaline soil. Lime also breaks down heavy clay soils, making them more porous and thus better drained and more workable. There are also important side effects, such as creating a better environment for bacterial processes, for example nitrification. The use of chalk or marl, which is a calcareous clay containing typically 60 per cent (but up to 90 per cent) calcium carbonate, can be traced back to prehistoric times. Where chalk was available it was spread in vast quantities, up to 50 tons per acre (125 tonnes per hectare), leaving the weather cycles of frost/warmth and rain/drought to break it down to a finer state. This is still practised in some southern and eastern areas of Britain. Sea-sand, in areas where it is highly calcareous, was used in large quantities as an alternative.

Solid calcium carbonate, however, is practically insoluble and only effective when in powdered form, which increases its surface area. Today, limestone is ground to a powder at an economic cost and, even allowing for transport and handling, is available to farmers at an acceptable price. In the past, a fine lime powder could only be obtained by burning limestone or chalk in kilns.

Chemistry of Lime

BURNING (CALCINING) at 900°C

limestone → quicklime

$$CaCO_3 \longrightarrow CaO + CO_2\uparrow$$

mol. wt.   100   56

SLAKING (HYDRATION)

quicklime + water → slaked lime

$$CaO + H_2O \longrightarrow Ca(OH)_2$$
$$\text{excess} \longrightarrow \text{lime putty}$$

REVERSION (HARDENING)

$$Ca(OH)_2 + CO_2 \longrightarrow CaCO_3 + H_2O$$

+ water + sand → mortar

Model of a limekiln at Mouth Mill, Devon, in Hartland Quay Museum. There are two arched draw-hole access recesses. The pot is about 10 feet (3 metres) in diameter at the rim and about 16 feet (4.9 metres) deep, narrowing to about 4 feet (1.2 metres) diameter at the base. On the left was a lean-to limeburner's hut with an oven for cooking meals, bleeding heat from the kiln. The shelter on the right could be used for storing burnt lime, tools and so on. Limestone and culm from South Wales were landed on the adjacent beach.

A rural flare kiln, from W. H. Pyne's 'Microcosm', 1808. The packhorses shown would bring limestone to the kiln and carry the burnt lime to the fields for slaking and spreading.

The quicklime drawn from the limekiln was *lump lime*, alternatively known as *lime shells*, and resembled the unburnt material in shape and size. To produce a powder suitable for spreading it had to be slaked. It was common practice for farmers to take lump lime straight from the kiln to the fields. This had its dangers, since the heat generated if it started to slake could set fire to carts or panniers, yet they would risk carrying it long distances. The lump lime would be dispersed in small heaps over a field, often covered with earth, and left to slake, in due course falling into a powder, which was then ploughed in. Alternatively, it was heaped along one side of a field and periodically turned over to assist slaking. This practice was observed in Wales as recently as the 1930s.

In the later nineteenth century, hydration plants were introduced. In these, burnt lime is ground and sprinkled with water, the resulting hydrated lime being dried and bagged for sale. Today, lime in this form is favoured for horticulture.

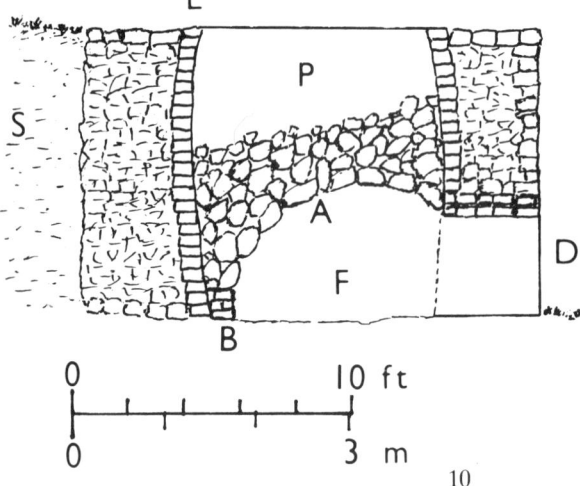

Features of a flare kiln. The shape and size drawn were typical of farmers' kilns in Surrey and Sussex. Variations occurred in other regions, and commercial flare kilns were larger. P, pot; L, lining; D, draw-hole; B, bench; A, arch or dome of material; F, fire cavity; S, natural slope or bank.

Soil regularly loses lime as it is washed out and as crops are lifted. The rate of loss varies from 1 to 8 cwt of equivalent calcium oxide per acre per year (125 kg to 1 tonne per hectare). Leaching occurs even in chalk and limestone regions, lime deficiency resulting in spite of the underlying calcium carbonate. The lime requirement varies enormously with soil type and local rainfall, but modern authorities quote 1 to 2 tons per acre (2.5 to 5 tonnes per hectare) applied as a rotational dressing every four to six years. For initial correction of acidity, far more is required, 3 to 5 tons per acre (7.5 to 12.5 tonnes per hectare) being common. In the past, even higher rates were used in some areas and some land today still benefits from the heavy liming of over a hundred years ago.

Chemically, 100 units of pure limestone yield 56 units of quicklime, but a ratio of 2:1 is usually adopted, assuming 90 per cent purity. Thus 2 tons of limestone must be burnt to produce 1 ton of quicklime. Contemporary literature of the nineteenth century quotes draw-kiln yields of 300 to 450 bushels per day. A bushel is a measure of volume but such yields would represent, say, 10 to 15 tons, enough to lime only 2 to 5 acres (1 to 2 hectares). This simple sum allows one to appreciate why so many kilns were built and why the invention of continuous-burning kilns was so significant. It also illustrates why large limestone quarries and chalkpits were opened specifically to meet the demand for agricultural lime.

KILN STRUCTURE

A limekiln of the form most commonly found until the advent of rotary and horizontal kilns, and not entirely obsolete, comprised an open-topped combustion chamber with one or more draw-holes or *eyes* at the base. Through these the fire was lit, the ashes were raked out and the burnt product was usually extracted. The combustion chamber is variously referred to as a crucible, bowl, or even well, but here the simplest term, pot, is used.

An *intermittent* or *periodic* kiln is loaded, fired, cooled and emptied, then reloaded for the next firing. A *continuous, perpetual, running* or *draw* kiln is kept burning and further supplies of raw material and fuel are fed in as the quicklime is drawn off. A *flare* kiln is one in which the fuel is not in contact with the charge of limestone or chalk, as distinct from a *mixed-feed* kiln.

The simplest and earliest type of stone-built kiln (as distinct from clamp kilns) was an intermittent flare kiln using wood or peat as fuel. An arch or dome of large lumps of limestone or chalk was built to form a cavity in which the fire was lit; the remainder of the charge was then added to fill the kiln. This was not economical but produced excellent lime unmixed with ash, especially if it was unloaded from the top when cool.

The earliest description of such a kiln is found in *De Agricultura* by Marcus Porcius Cato, a Roman who lived from 234 to 149 BC. He described it as 20 feet (6.1 metres) deep and 10 feet (3 metres) in diameter, narrowing to 3 feet (0.9 metres) at the top. He suggested a pit to hold the ashes, with the proviso that if two entrances were made the pit was unnecessary, the ashes being cleared through one entrance while the fire was in the other. He mentioned a *fortax* round the base, usually translated as a ledge on which the initial dome of limestone rested. Cato also stressed the importance of minimising the draught and stopping up holes in the structure with clay. Although written so long ago, Cato's description of a flare kiln would have been recognisable to a limeburner two thousand years later. The need for slow steady burning and the suggestion of two draw-holes or stoke-holes are particularly appropriate.

Excavated Roman limekilns generally have similar diameters to Cato's but they are often squatter, as little as 8 to 10 feet (2.5 to 3 metres) deep and cylindrical, with single stoke-holes. The internal ledge is found, and sometimes an ash-pit. The burnt lime may have been unloaded from the top, to avoid contamination with ash, though the presence of a raking-

Seventeenth-century flare kiln at Trim Castle, County Meath, Republic of Ireland. It has a cylindrical pot just under 10 feet (3 metres) in diameter and 5 feet (1.5 metres) deep, with two draw-holes.

out pit in some cases would indicate the drawing of lime through the stoke-hole as generally done in later kilns. A battery of six kilns excavated at the Roman legionary fortress at Iversheim, Germany, had walled stoking areas for weather protection and additional structural strength. The fuel was wood selected to give a fierce heat initially, followed by harder wood such as oak to maintain the temperature for several days. The fire could be at the inside end of the long stoke-hole to heat the charge on the muffle principle. The complete cycle of loading, firing, cooling and extraction may have taken up to a fortnight.

Medieval kilns sometimes resembled Roman ones but some were simply unlined or clay-lined pits with stoke channels cut to the base. They could be very small, say 4 to 5 feet (1.2 to 1.5 metres) wide at the rim and about 6 feet (1.8 metres) deep. Stoking tunnels, where present, could be of brick, tile or local stone as available. Some much larger ones have been excavated, however: one at Southampton had a pot 13 feet 6 inches (4.1 metres) in diameter at the rim, narrowing to 3 feet (0.9 metres) at the base, with a stepped ramp down to an arched draw-hole and three other vents or stoke-holes.

At medieval castles, as might be expected, more substantial limekilns were built for larger quantities of lime for mortar-making. An example at Ogmore Castle, Mid Glamorgan, had thick drystone walls enclosing a pot of tapering section. It had two splayed draw-holes and was coal-burning. Such kilns would have been fired intermittently to make sufficient lime for a particular phase of building and then abandoned.

Small intermittent kilns or *field* kilns were a common sight on or near farmland in the eighteenth and nineteenth centuries but originated in the sixteenth. George Owen of Pembrokeshire (1552-1613) described a kiln 6 feet (1.8 metres) high, having a pot 4 to 5 feet (1.2 to 1.5 metres) broad at the rim and narrower at the base, with two eyes. It was fired with a mixture of coal or culm and wood, but field kilns in areas where wood was plentiful used brushwood, coppiced wood or furze as fuel.

In Surrey and Sussex, a typical field kiln had thick sandstone walls with brick used for lining the pot and for constructing a front wall with a single arched draw-hole. Buttresses of sandstone and brick strengthened the wall against internal pressures where necessary. The pot was almost vertical-sided, narrowing only slightly at the rim and the base, about 8 to 10 feet (2.4 to 3 metres) in diameter and of similar depth. A ledge or bench ran around the inside; with the help of a wooden frame or iron horse the initial load of chalk was formed into a dome resting on this ledge. Smaller pieces of chalk were then loaded to fill the kiln. A fire was lit under the dome and modest heat maintained initially to dry and set the charge, then fierce heat was applied until calcining was complete in 24 to 36 hours, indicated by a clear red fire at the top. After a lengthy cooling period, the lime was drawn out. Long-handled tools were used for raking and clearing ashes. The complete operation probably took four or five days.

Such flare kilns were built into slopes or roadside banks. No weather protection is apparent at surviving structures although a simple shelter may have been erected. The draw-hole was partly blocked with earth and stones to reduce the draught during calcining. The kiln would need continual repair and new kilns were sometimes erected, re-using materials as far as possible. Abandoned kiln sites can be identified from debris even where the hollow indicating the pot has been filled in. Field names such as Kiln Field or Old Kiln Field are other indicators and Furze Field recollects a plantation of furze created for fuel.

Field kilns in other English counties were similar in size but the shape of pot and draw-hole varied to suit the local building materials. Many survive in the Peak District and the Yorkshire Dales, for example. The author has also observed small field kilns in Ireland which probably used peat as fuel.

For seasonal production of agricultural lime, in some areas limestone was alternatively burnt in clamps. A primitive form known as a *sow* kiln was used in Northumberland even into the nineteenth century. A bowl was scooped out of a hillside and limestone laid on a bed of kindling wood. Further layers of limestone with wood or coal were added, then the whole was covered with turves to ensure slow burning, leaving a flue for lighting and a vent hole in the top.

Above: *Model of a Surrey farmer's kiln in Guildford Museum. The internal bench or ledge, on which the initial load of chalk was rested to form a dome, is visible at the base of the pot.*

Below: *Draw-hole and front face of a brick-built Surrey farmer's flare kiln. In an area lacking hard building stone, brick was used for the arched draw-hole and hard stock brick for the pot lining. However, the local sandstone could be used for thick insulating walls and sometimes in buttresses to support the front face.*

Small farmer's kiln beside the river Swale, North Yorkshire. It is built of thick dry-stone walls for good insulation, using the plentiful local limestone. It could easily be repaired or rebuilt by craftsmen familiar with building dry-stone walls and probably represents a design unchanged for many years.

Stoke-hole of a farmer's kiln at Monyash, Derbyshire, last fired in about 1800. There is a steep drop to the fire opening at the base of the pot, so the burnt lime may have been extracted from the top. The flowerpot-shaped pot varies in diameter from about 8 feet (2.4 metres) at the rim to 5 feet (1.5 metres) at the base and is about 8 feet (2.4 metres) deep. There is a model of it in the Derby Industrial Museum.

A much larger type of clamp kiln was used in Peeblesshire, Borders, Scotland, paralleled by the *pye* or *pudding-pye* kilns of Derbyshire. They were 40 to 50 feet (12.3 to 15.3 metres) long and up to 20 feet (6.1 metres) wide, built up with alternate layers of limestone and fuel. The pudding-pye kiln at Newhaven, Derbyshire, described by John Farey in 1811, was built against a bank and rose to a height of 14 feet (4.3 metres), including the heaped top. It had a roughly built stone front wall with draw-holes but this was partly dismantled to extract all the lime. It took up to twelve days to burn and cool but produced as much as 80 tons of lime at one firing.

The *horseshoe* kilns at Dudley, West Midlands, were a type of mixed-feed intermittent kiln similar to a clamp in principle but lined with brick or stone and provided with a front wall broken down after firing.

The small coastal kilns depicted in old drawings and paintings have generally vanished. The single kilns still standing today at small harbours and beaches are cylindrical or square stone blocks, containing tapered pots and one or two draw-holes. Many are unlined and were probably fired intermittently, fuelled with culm. Some were described as running kilns but perhaps only operated as such over a limited period. They generally date from the nineteenth century, replacing the older, smaller kilns. Harbour kilns were warm places at which to meet for a gossip. Children would bake potatoes on them and daring boys would jump over the hot kiln tops!

The continuous draw kiln came in about 1750 as the increasing demand for

A small flare kiln at Kinvara, County Galway, Republic of Ireland. (Above) It has a square draw-hole and stepped stone lintels reducing the height to the fire opening. (Below) The pot is made of roughly cut local stone.

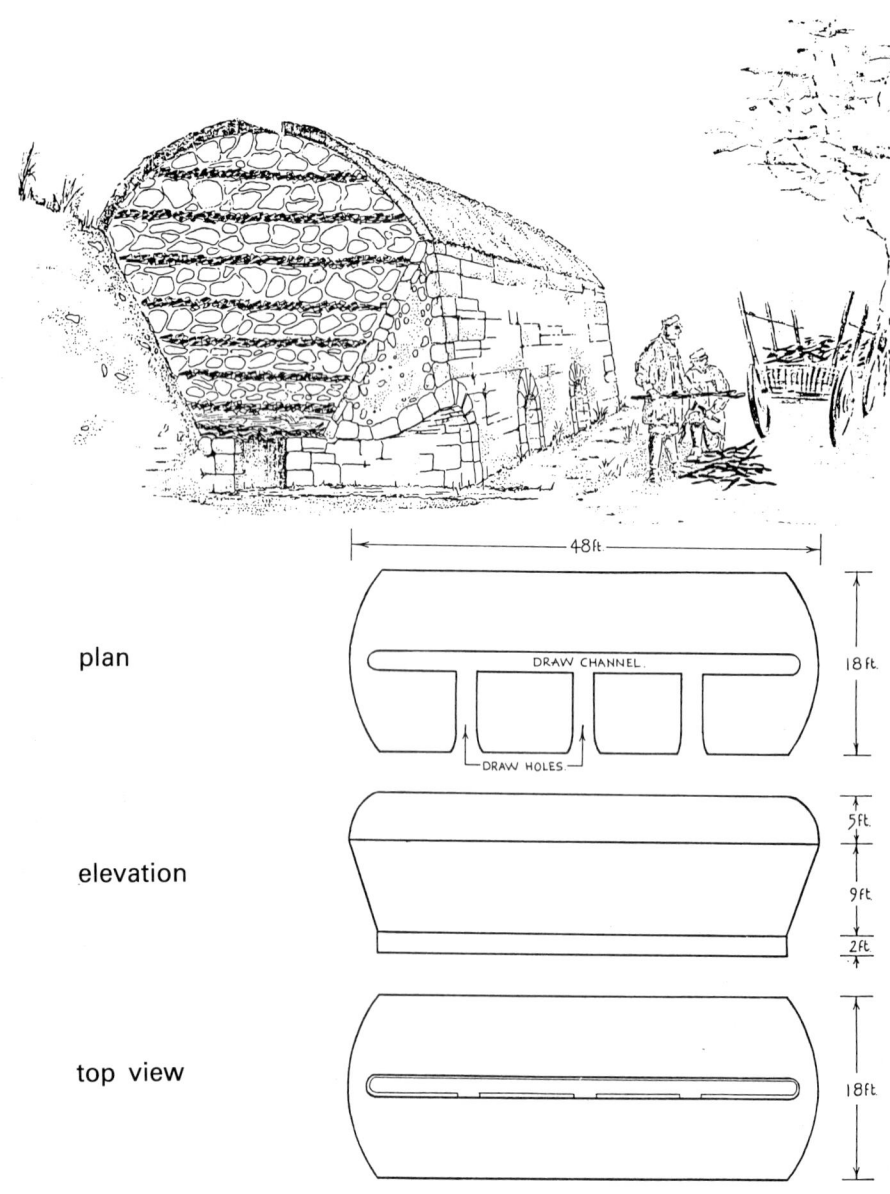

A 'pye' or 'pudding-pye' kiln as used at Newhaven, Derbyshire, in the early nineteenth century, drawn from a description by John Farey, 1811. It was filled with alternate layers of limestone and coal, covered with turves. The fire was started with brushwood, heather and straw laid in the draw channel and lit through the draw-holes. It was kept burning for five to ten days to produce about 80 tons of lime.

Limekiln at Combe Martin, Devon, in about 1824. A woman is hanging out washing to dry on the rim of the kiln. Note the steps to the top and the tall draw-hole arch. From a watercolour by J. M. W. Turner at the Ashmolean Museum, Oxford.

lime required higher output, greater efficiency and longer life. The majority of the disused large kilns at quarries, harbours and estuaries are of this type, although flare kilns of improved design continued in use for production of good-quality lime. Typical draw kilns have pots lined with hard stone or firebrick capable of withstanding continuous firing over a long period. Their profile resembles an inverted cone or is shaped like an egg standing on its narrow end with the top and bottom sliced off. Some are cylindrical for about two-thirds of their depth, narrowing sharply at the bottom and turned in slightly at the top. Many theories were advanced for the ideal shape, not all of them tenable, but the chief aims were to maintain the calcining temperature with fuel economy and to achieve good lime flow without *hang-up*, that is, failure of the burnt lime to settle for drawing off at the base. They were loaded with alternate layers of limestone and coal in the ratio of three, four or five parts of limestone to one of coal. Movable iron bars at the base of the pot provided support for the charge and could be adjusted for drawing off the lime, but alternative arrangements are found including chutes to the draw-holes built into the structure.

They have from one to four draw-holes in arched recesses in the thick kiln walls, showing local design variations reflecting building techniques to suit the available stone. For example, pointed arches formed by corbelling are found in some areas and rounded arches in Romanesque style in others. Draw-hole access recesses are high enough to allow a man to stand upright at the entrance but the sides and roof are tapered or stepped down towards the small draw-hole cum stokehole. Sometimes a *poking-hole* is found above a draw-hole, for clearing ash and encouraging the burnt lime to settle. At the initial lighting, the fire spread progressively through the kiln; it could be loaded gradually as the fire spread. As quicklime was drawn off, fresh layers of limestone and coal were loaded into the top. Inspection doors are sometimes found for testing the temperature and checking on the progress of firing.

For good insulation and structural strength, kiln pots were formed in stone blocks, often in pairs, with walls of

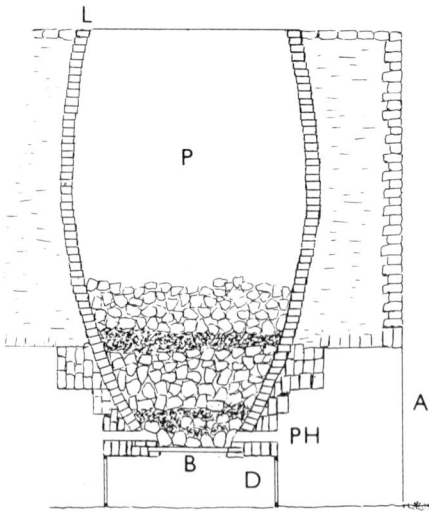

Left: *Features of a draw kiln, showing limestone and coal in alternate layers. Pot diameters varied from 8 to 16 feet (2.4 to 4.9 metres) and the depth varied from 20 to 30 feet (6.0 to 9.1 metres). The shape varied in profile, but generally narrowed to 3 feet (0.9 metres) or less at the base and usually curved in at the top. Up to four draw-holes are found. P, kiln pot; L, lining; D, draw-hole; A, access arch; PH, poking-hole; B, support bars (a separate grate is sometimes also found). The kiln is shown partly loaded.*

Below: *Rim of a draw kiln, showing the lining of header bricks.*

Bottom: *Appearance of the top of a draw kiln when filled; only a layer of slack coal on top is visible and there is nothing to prevent a person from walking into the hot kiln!*

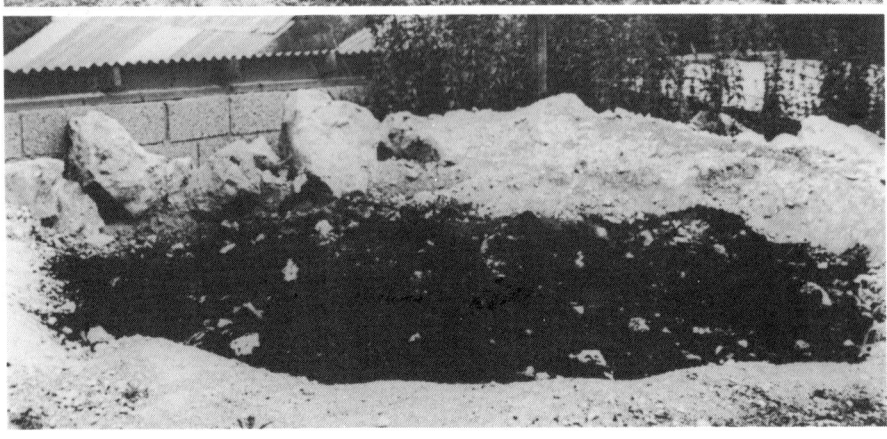

Above: *Interior of a draw-kiln pot, showing how it narrows towards the base.*

Right: *Derelict draw-hole of a draw kiln exposing the bars on which the initial blocks of limestone rested; such bars were adjustable. This one has a fire grate below so that the ash could be cleared.*

Below: *Large circular single kiln near Thorngrafton, Northumberland, with multiple draw-holes. Note the tall access recesses formed by a corbelling technique and the massive blocks of the lower courses. A long earth ramp behind leads to the kiln top.*

One of the draw-hole recesses of a limekiln at Wallog, Dyfed, showing the well built Romanesque arch and stepped arches over the draw-hole itself, which is just visible though blocked by debris.

Single kiln with a pointed-arch draw-hole recess at Mwnt, near Cardigan, Dyfed (National Trust). It is built into a bank at the top of a cliff overlooking the small beach. Such kilns were very numerous in south-west Wales.

Block of two kilns at Lynmouth, Devon, with a central access lobby serving both kilns. The external draw-hole arches have been preserved to form useful shelters.

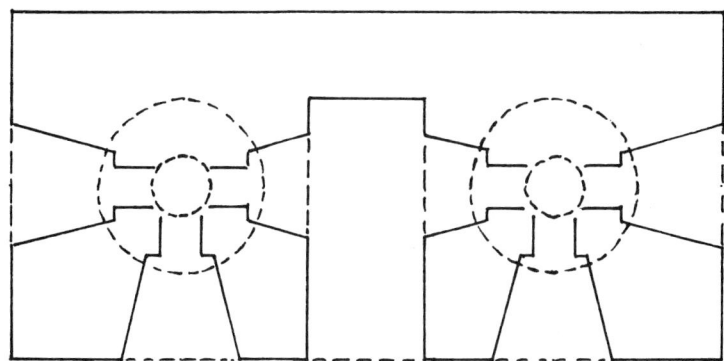

Plan view of one arrangement of a double-kiln block. Many variations are found, depending on the number of draw-holes. This arrangement sometimes occurs in blocks of three or four kilns.

coursed masonry infilled with rubble or sand, at least 3 feet (0.9 metres) thick at the narrowest point and thus much thicker at the base. They were generally built against low cliffs or banks, giving easy access to the charging level. A central tunnel-like lobby is often found, giving access to each kiln of a pair and providing cover for tools, barrows, bags and kindling wood.

Blocks or batteries containing several kilns were common at quarries, beside inland waterways and at some estuaries and harbours. Some are of elaborate construction, with covered passages between kilns; in the larger ones, carts and even rail wagons could be taken inside a block. Good examples of kiln blocks exhibiting these features can be seen at Cotehele Quay and at Morwellham Quay, both on the banks of the Tamar on the Devon/Cornwall boundary. They date from between 1770 and 1830. At Beadnell harbour, Northumberland, a complex and architecturally pleasing arrangement of four kilns with multiple draw-holes and connecting tunnels, partly intact, dates from the late eighteenth century. Interesting kilns are found in Norfolk with features peculiar to the

Block of three limekilns at Cotehele Quay, Cornwall, on the Tamar river (National Trust). The half-arch access tunnels are a feature of other kilns in the area, indicating a local building technique.

Lobby giving access to two kilns in a block at Morwellham, Devon, on the Tamar. The narrow arch at the rear leads to a passage providing a covered way for the limeburners between kilns. The blocked-up arch to the left indicates some rebuilding — a frequent necessity in blocks remaining in use for several decades.

area; an underground circular tunnel, into which wagons could be taken, runs around a wide brick column inside which are chutes descending from the base of the pot. At an intact one at Coltishall, the central column flares out to meet the tunnel roof, forming an example of vaulting which has caused it to be dubbed a *cathedral* kiln. It would do credit to any architect. Another sophisticated design can be seen on Holy Island, Northumberland. Here there are six kilns in a large block, with tunnels allowing rail wagons to serve the draw-holes.

Although draw kilns of the basic design remained in use into the twentieth century, there were many attempts at improvement, not all successful. Stuart Menteath of Closeburn, Dumfriesshire, described in 1813 a kiln with an oval pot having a brick lid fitted with three chimneys and three pairs of hinged double iron doors through which limestone and coal were loaded. At the base were three pairs of iron doors, each comprising a lower door for removal of ash and small lime and an upper door for drawing off lump lime. The lid and chimneys were claimed to provide weather protection, to control the draught and to allow the kiln to be shut off with the contents kept hot over a period of low demand, being relit after a day or two. Menteath further

A sophisticated block of six kilns on Holy Island, Northumberland (National Trust). Internal tunnels, large enough to accommodate tubs on rails, gave access to the draw-holes. In each kiln the burnt lime was directed to the multiple draw-holes by chutes built into the base of the pot. Horse-drawn wagonways connected the kiln block with a limestone quarry on the island and with a quay; coal was brought in by sea. The complex dates from the 1860s, replacing older kilns.

Flare kilns with bottleneck-shaped chimneys at Dorking chalkpits, Surrey, from a pen-and-wash drawing by G. Scharf, 1823. The burning kiln's draw-hole is roughly blocked to reduce the draught. Loading doors are visible in the chimneys. Note also the lean-to shed on the right. A larger open-topped kiln is being fired on the left of the picture.

claimed that the closed top would prevent a kiln from being mistaken for a coal-fired beacon if built on the coast, as had happened in 1810 when a ship was wrecked off the Isle of May, Fife, as a result of such confusion!

Domed tops or chimneys were sometimes built over kiln pots for draught control. Where this was done a loading door was necessary for charging, sometimes with a smaller iron door above, opened or shut as required for temperature control. Flare kilns with bottleneck-shaped chimneys were used at chalkpits in Dorking, Surrey. Kilns with domed tops were in use at Merstham, Surrey. However, kilns with tall domed or bottle-shaped tops were generally cement kilns.

James Malcolm in 1805 designed a double *rectangular* kiln for chalk burning. The heat of one kiln, fired first, would pass through a party wall to the other to dry its load and hence reduce its burning time when fired in its turn. John C. Morton in 1855 described a rectangular intermittent kiln similar to a brick kiln. At one end were two fireplaces closed with iron doors and at the other a perforated brick wall leading to a fire chamber and a flue and chimney. It was loaded with chalk or limestone, mixed with slack coal near the bottom and burnt for two or three days, then sealed up for another couple of days before drawing. Remains of such kilns exist in Surrey and Sussex but are rare.

Count Rumford, founder of the Royal Institution, designed a kiln in which fuel and charge were kept separate, yet both could be added continually, to form, in effect, a perpetual flare kiln giving ash-free lime. It was ingenious but too complicated for practical use, but eighty years later vertical shaft kilns exhibiting some of the features of Count Rumford's kiln were being operated. An example was the Dietzch kiln, designed for cement burning but converted for limeburning at Betchworth, Surrey, in 1887. They were built in pairs inside tall brick towers. Chalk descended through a pre-heating zone to a firing chamber where the fuel was fed in, then passed through a cooling zone to a grate at which the quicklime was extracted.

Left: *Vertical shaft kiln built by Count Rumford in Dublin in about 1800. The shaft, with brick cavity walls, was 15 feet (4.6 metres) tall, 9 inches (0.23 metres) in diameter at the top and 2 feet (0.61 metres) in diameter at the base. Fuel was injected through a hinged iron door at point A and kept separate from the limestone, which descended through a drying and pre-heating zone into a burning zone, followed by a cooling zone. The lime drawn off was free of ash. A door at B allowed clearing of the flue and there was an ash-pit door at C. In effect it was a continuous flare kiln. (From Malcolm's 'Compendium of Modern Husbandry', 1805.)*

Right: *Simplified drawing of a Dietzch cement kiln modified for quicklime production, at Betchworth, Surrey, in 1887. It was very tall and chalk was loaded at the top into a pre-heating zone. Fuel was fed in at the shoulder and the chalk raked to join it in the burning zone just below. The burnt lime descended through a cooling zone to the draw-hole. These kilns were always built in back-to-back pairs. They were used at several American cement plants.*

At nearby Brockham, Surrey, Alfred Bishop patented the 'Brockham kiln' in 1889. The pot was only 4 feet (1.2 metres) in diameter at the rim, narrowing towards the draw-hole. A set of curved iron chutes, typically seven, were sunk into the ground in a circle concentric with the pot, into which coal dust was fed to emerge in the burning zone of the kiln. Above the pot was erected a roughly conical tower with a large loading door into which chalk and coke were loaded. Chalk was dried by the rising heat as it descended to the burning zone. It was very labour-intensive and not many were built. The Aalborg kiln was similar in principle.

Also at Betchworth, Hoffmann kilns,

Above: *Dietzch kilns at Betchworth, Surrey. The fuel-feed level is where the corrugated iron is visible. Chalk was carried by aerial ropeway and remains of the loading gantry can be seen.*

Right: *Top structure of a Brockham kiln showing the chalk-loading door.*

patented in 1858 for brickmaking, were adapted for limeburning. This type of kiln comprised a series of firing chambers or unpartitioned cells in a circular or oval tunnel with openings to a central flue and chimney. The fire progressed slowly around the tunnel so that between twelve and twenty chambers were fired in sequence. The operations of loading, cooling and clearing through draw-holes were thus similarly sequential. A Hoffmann limekiln at Buxton, Derbyshire, was operated from 1872 until 1944; one at Langcliffe Quarry, Settle, North Yorkshire, used from 1873 to 1939, is substantially intact and may be restored. The limestone had to be hand-stacked and small coal fed in manually from above, so Hoffmann kilns were very labour-intensive. A block of De Wit kilns of Belgian design, built at Amberley, West Sussex, in about 1905, bore a similarity to the Hoffmann. Eighteen chambers were fired in sequence, but they were arranged in line in back-to-back pairs rather than

A Hoffmann kiln in operation at Harpur Hill, Buxton, Derbyshire, in 1930. It has now been demolished.

Below: Men loading the Hoffmann kiln at Harpur Hill.

Trackway leading from the beach at Porthysgadan, Lleyn Peninsula, Gwynedd, to the limekiln on the cliff top at the left of the picture. Tracks serving both the charging and draw-hole levels are clearly visible.

in a closed loop. It was unsuccessful and they had been converted to draw kilns by 1911.

Steel rotary kilns, designed in 1885 for Portland cement, are found at present-day lime-production plants. They are very long, slowly rotating cylinders on a slight slope and have pre-heating, burning and cooling zones. They may be fired by oil, gas or solid fuel. Another type is a horizontal tunnel kiln in which trucks carry lime through the zones as in a brickmaking tunnel kiln. Modern designs of vertical shaft kilns are also used for producing ash-free lime. Details of these are beyond the scope of this book.

TRANSPORT AND LOADING ARRANGEMENTS

In the past, water transport was of prime importance in getting raw material and fuel to limekilns. Coal was exported from Newcastle to the Thames as early as the thirteenth century. Coastal kilns in the South-west obtained fuel and limestone by sea from South Wales. Even where limestone occurred locally, as at Plymouth, Devon, fuel still had to be imported. Elsewhere the story was similar: burnt lime was sometimes exported by sea and British coal was exported to such far-away places as the Canary Islands, where disused limekilns can be seen.

Sailing barges were commonly used on the Thames, the Tamar and elsewhere. A major coastal trade sprang up as a result of limeburning. For example, in the 1860s fifteen ships regularly brought in limestone for kilns in Aberystwyth harbour, Dyfed. At smaller harbours and coves, ships would be beached and the cargo of limestone or fuel carried in baskets, by pack animals or carts, to the top of a nearby kiln.

Overland carriage was by packhorse or carts. Sometimes the burnt lime was carried long distances to fields; farmers would travel at night to a coastal kiln and queue up at dawn to get the first lime drawn, then trudge back by day. Sleds were used in some areas for short distances from kiln to field. In contrast, inland wood-burning kilns in Surrey, for example, were built on the farms and the raw material carted from chalkpits several miles away. Where turnpike roads were used, there was often exemption from tolls, or a reduction, for lime used as a fertiliser.

The advent of canals allowed limestone and fuel to be distributed far more widely and canals were cut specifically for this purpose. For example, the lime trade was the prime motive for building the Montgomeryshire Canal, Powys, in 1796/7. In hilly country, this canal required aqueducts, tunnels and a large number of locks; by 1821 its total length was 34 miles (54.7 km). Kiln blocks were built alongside wharves as the canal was made; by 1840-1 there were 92 limekilns along a 26 mile (42 km) stretch and a peak carriage of 56,501 tons of limestone per annum was achieved. Another example was the Rolle Canal on the Torridge, Devon, cut to extend the carriage of limestone and coal extracted in South Wales further up river from Bideford to reach Little Torrington, at a cost to Lord John Rolle of between £40,000 and £45,000. Although only 8 miles (13 km) long, this waterway was a significant improvement over expensive land carriage by poor roads, allowing inland farms the benefit of liming at an economic price.

At Dudley, West Midlands, canals were built inside the hills to bring out excavated limestone, required as a flux for blast furnaces as well as for limekilns.

The building of the great docks and other projects in London in the early nineteenth century led to expansion of chalk extraction in Surrey and Kent, using the Medway and Thames from Kent and the Wey Navigation from Guildford, Surrey, to convey chalk to limekilns at the dock sites, and taking coal as a back carriage for local limeburning. The Wey Navigation tolls and the poor quality of the lime made it difficult for Guildford to compete with Kent, however, and the Surrey industry moved to Merstham.

The Surrey Iron Railway, opened in 1803 as a plateway for horse-drawn

Lord Ward's limekilns at Dudley, West Midlands, now in the Black Country Museum. On the left is part of the original block of three draw kilns built in 1842; the one on the right was added later. The loading crane on rails dates from as early as 1910. Limestone and coal were hauled up through shafts in the kiln block.

Mixed-feed kilns at Perseverance Works, Peak Dale, Derbyshire, served by a siding from the Peak Forest Tramway. They were operated from 1847 to 1939 and have now been demolished.

wagons between the Thames and Croydon, was extended in 1805 to the Merstham stone quarries and chalkpits. It was the first public railway sanctioned by Act of Parliament independent of a canal. The loads which could be horse-drawn on rails, compared with haulage on rutted and muddy roads, made it a revolution in transport.

Tramways had previously been laid in conjunction with waterways elsewhere. An example was the Peak Forest Tramway, Derbyshire, authorised by the Peak Forest Canal Act of 1794 and completed by 1799 to carry limestone from the Doveholes quarries to the Ashton-under-Lyne Canal. Horse-drawn wagons ran on an L-section plateway and there was a gravity-worked incline south of Chapel-en-le-Frith.

With the coming of steam railways, new limekilns sprang up served by branch lines or sidings, and railway companies entered the limeburning business. An example was the Stanhope and Tyne Railway Company in County Durham, which built kilns in 1834 and connected the limestone quarries and kilns at Stanhope with South Shields using a mixture of horse power, self-acting inclines, stationary engines and locomotives. Ten years later, the Stockton and Darlington Railway Company built much larger oval kilns and laid rails to the kiln tops for increased production. Similar schemes existed in other parts of Britain. However, expansion of the railway network led to abandonment of vast numbers of canalside kilns and field kilns. It was poor economics to transport 2 tons of limestone to make 1 ton of quicklime when the latter could be produced in bulk at quarries and delivered by rail more cheaply. Road transport has now replaced rail.

Limestone was broken up on the kiln top and loaded manually but no protective rails were provided. There are many recorded instances of men falling into kilns with fatal results, such as tramps sleeping too near the rim of a hot kiln. Pyne's *Microcosm* relates one story of a Lincolnshire nobleman who, on passing a kiln, heard a piercing shriek; he galloped

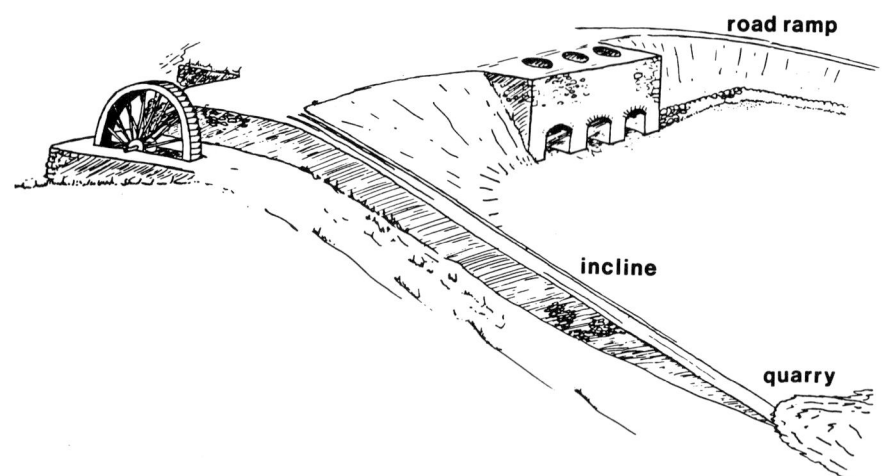

Water power was used at Park, Closeburn, Dumfriesshire, to haul tubs up an incline from a limestone quarry below the limekiln level. The coal came by road up the ramp on the other side of the kiln block. The scheme dates from about 1800.

up but was too late to save a limeburner from falling into the burning lime. Steps were sometimes built to allow the limeburner to move easily from draw-hole to charging level. At large draw-kiln blocks, a ramp was made to the kiln top where it was necessary to allow carts or wagons to unload straight into the kiln. As kilns became more mechanised horse-gins and, later on, steam engines, were installed to facilitate loading.

In some cases, water power was harnessed; good examples were at Closeburn, Dumfriesshire, and Moorswater, near Liskeard, Cornwall. At the former site, the limestone quarry is below the kiln level and an incline was built up to the kiln top. A large wheel was driven by water brought several miles by aqueduct and also powered a sawmill. The wheel was geared to a winch which hauled up loaded tubs assisted by empty

An unusual water-powered mechanism was used to haul materials to the top of a kiln block at Moorswater, Liskeard, Cornwall. An undershot wheel driven by the East Looe river was geared to a winch from which a cable passed up through a hole in the block to pull wagons up an incline. Details of the drive and braking mechanisms are shown.

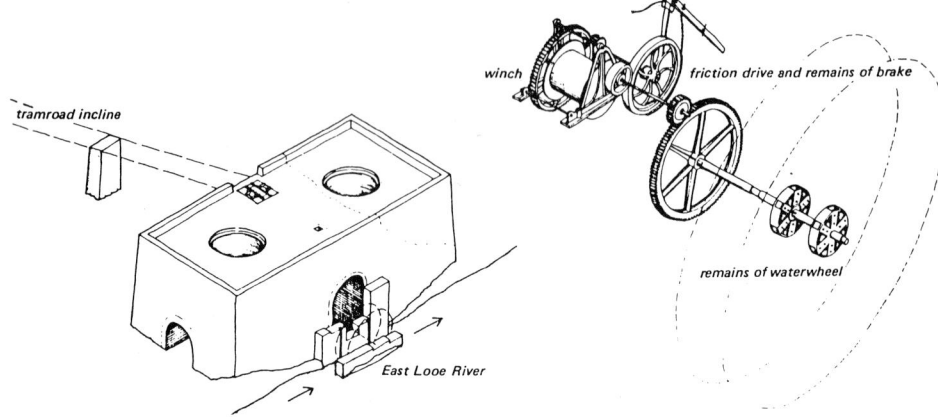

tubs descending under gravity. This scheme was devised by the Earl of Menteath in about 1800. At Moorswater, a waterwheel drove a winch housed at the rear of the high draw-hole access tunnel. The cable passed up through the kiln block to haul wagons up an incline. Loading was facilitated by provision of a turntable on the top of the kiln. Where kilns were sited downhill from a quarry or chalkpit, tramways and, later, narrow-gauge railways conveyed material to the kilns; ropeways were also used. Gantries and cranes assisted loading at some limeworks and railway lines served draw-holes, with platforms to allow lime to be loaded easily into trucks. Many remains, generally derelict, survive of these various arrangements.

FURTHER READING

There is no general work in print dealing exclusively with limeburning, although regional studies have been published. The following are examples of books and papers containing useful information or covering specific studies. Some may be difficult to obtain through normal public libraries, however.

GENERAL
Davey, Norman. 'Limes and Cements' in *History of Building Materials*, chapter 12. Phoenix, 1961.
Dix, Brian. 'The Manufacture of Lime and its Uses in the Western Roman Provinces', *Oxford Journal of Archaeology*, 1, number 3 (November 1982), 331-45.
Gardner and Garner. *Use of Lime in British Agriculture*. Farmer and Stockbreeder Publications, 1957.
Hudson, K. *Building Materials*. Industrial Archaeology Series, number 9. Longman, 1972.
Salzman, L. F. *English Industries of the Middle Ages*. Clarendon Press, 1923.
Searle, A. B. *Limestone and Its Products*. Ernest Benn, 1935.

REGIONAL
Many of the *Industrial Archaeology* series published by David and Charles, 1971-5, contain chapters on limekilns and lists of sites. The volumes covering *North-east England* (Frank Atkinson), *The Tamar Valley* (Frank Booker) and *The Peak District* (Helen Harris) are particularly useful.
Aldsworth, F. *Limeburning and the Amberley Chalkpits*. West Sussex County Council, 1979.
Havinden, Michael. 'Lime as a Means of Agricultural Improvement: the Devon Example' in C. W. Chalklin and M. A. Havinden (editors), *Rural Change and Urban Growth*, chapter 5. Longmans, 1974.
Holt, Margaret. 'Limekilns in Central Sussex', *Sussex Industrial History*, 2 (1971), 23-30.
Hughes, Stephen. *The Industrial Archaeology of the Montgomeryshire Canal*. National Monuments Record for Wales, 1983.
Mawson, D. J. W. 'Agricultural Limeburning: the Netherby Example', *Transactions of the Cumberland and Westmorland Antiquarian and Archaeological Society*, 80 (1980), 137-52.
Robinson, D. J., and Cooke, R. U. 'Limekilns in Surrey: a Reconstruction of a Rural Industry', *Surrey Archaeological Collections*, LIX (1962), 19-26.
Skinner, B. C. *The Lime Industry of the Lothians*. EUEA Studies in Local History, University of Edinburgh, 1969.

Board of Agriculture Surveys of the early nineteenth century have interesting contemporary comments on the use of lime and some describe limekilns. They were published on a county basis, and local libraries sometimes hold copies. Some nineteenth-century *Cyclopedia*, for example those of Morton, Rees or Wilson, are useful although not easy to find. Many include limekiln designs.

PLACES TO VISIT

Listed below are places where limekilns and/or limeburning exhibits are accessible to the public. The list is not exhaustive and all over Britain limekilns will be encountered, some in the care of public or private bodies, others simply standing unused and unprotected. Where appropriate, intending visitors are advised to find out opening times before making a special journey.

Amberley Chalk Pits Museum, Houghton Bridge, Amberley, Arundel, West Sussex BN18 9LT. Telephone: 079881 370.
Beadnell Harbour, Chathill, Northumberland (NT).
Belan Limekilns, Limekiln Lane, Welshpool, Powys. Enquiries to: Powys County Council, 2 Canal Yard, Severn Street, Welshpool, Powys. Telephone: 0938 4348.
The Black Country Museum, Tipton Road, Dudley, West Midlands DY1 4SQ. Telephone: 021-557 9643.
Buttington Wharf, Welshpool, Powys. Enquiries as for Belan Limekilns.
Cilgerran Castle, Cardigan, Dyfed (CADW). Excavated medieval limekiln.
Cotehele Quay, St Dominick, Saltash, Cornwall PL12 6TA (NT). Telephone: 0579 50830.
Derby Industrial Museum, The Silk Mill, off Full Street, Derby DE1 3AR. Telephone: 0332 293111, extension 740.
Hartland Quay Museum, Hartland Quay, Bideford, Devon EX39 6DU. Telephone: 028883 353. Contains a coastal limekiln model.
Holy Island, Northumberland (NT).
Morwellham Quay Open Air Museum, Morwellham, Tavistock, Devon PL19 8JL. Telephone: 0822 832766.
National Stone Centre, Wirksworth, Derbyshire DE4 4FR. Telephone: 0629 824833.
Ogmore Castle, Bridgend, Mid Glamorgan (CADW). Excavated medieval limekiln.

Rail tubs were used to simplify the loading of limestone into a kiln pot at Cowdale, near Buxton, Derbyshire.